ECLIPSE MA

2024's Upcoming Solar Eclipse Is a Must-See Event

By Evan B. Sherill

COPYRIGHT

All rights reserved. No part of this publication may be reproduced, distributed, or transmitted in any form or by any means, including photocopying, recording, or other electronic or mechanical methods, without the prior written permission of the publisher except in the case of brief quotations embodied in critical reviews and certain other non-commercial uses permitted by copyright law.

Copyright © Evan B. Sherrill, 2023

TABLE OF CONTENTS

Introduction to Solar Eclipse

2024 Solar Eclipse: Celestial wonder, thrilling astronomers

Best Viewing Spots For 2024 Solar Eclipse

How Improved Technology Will Enhance The Upcoming Eclipse Viewing Experience

Types of Solar Eclipses and their differences from each other

Unique Challenges Associated with Viewing of Solar Eclipses and How to Overcome Such

How can you capture the best photos and videos of the 2024 Solar Eclipse, and what gear should you bring with you?

Some Cultural and Historical Significance of Solar Eclipses and The various Responses of The Populace Towards Such Over The Years

Introduction to Solar Eclipse

A solar eclipse is a celestial event that occurs when the Moon passes between the Sun and the Earth, partially or completely blocking the view of the Sun. The phenomenon is the result of a rare alignment of the three celestial bodies and occurs about two to five times a year, with each eclipse visible from only a narrow strip of the Earth's surface. Solar eclipses have been observed and recorded by human civilizations for thousands of years and have held significant cultural and scientific significance throughout history.

Solar eclipses come in two primary forms: **total and partial eclipses.** A total solar eclipse occurs when the Moon completely covers the Sun, and only the Sun's corona (its outer atmosphere) is visible from Earth. The corona is a halo of plasma that surrounds the Sun and is only visible during a total solar eclipse. The total eclipse typically lasts for only a few minutes, with the exact duration depending on the location of the observer.

A partial solar eclipse occurs when the Moon only partially covers the Sun. During a partial eclipse, the Sun appears as a crescent, and the amount of the Sun covered by the Moon depends on the observer's location. Partial eclipses can last for several hours.

Solar eclipses have been significant to human cultures for thousands of years. The ancient civilizations of China, Greece, and Egypt all recorded eclipses in their histories and used them to make astronomical predictions. In many cultures, solar eclipses were seen as omens of bad luck or foretellings of disasters. In some cultures, the eclipses were believed to be the work of dragons or other supernatural beings who were "eating" the Sun.

Solar eclipses have also been important for scientific study. In the 20th century, scientists used eclipses to study the Sun's corona, which is not visible during normal daylight hours. By studying the corona, scientists were able to learn more about the Sun's magnetic field, temperature, and other important characteristics.

In recent years, the study of solar eclipses has become more accessible to amateur astronomers and the general public. The advent of digital cameras and the internet has made it possible for people all over the world to capture and share images of eclipses. The increased interest in solar eclipses has led to more organized viewing events and eclipse-related scientific studies.

It is important to note that viewing a solar eclipse without proper protection can cause serious eye damage. Staring directly at the Sun, even during an eclipse, can damage the retina and cause permanent vision loss. To safely view a solar eclipse, one should use special solar filters or eclipse glasses, or indirectly view the eclipse using a pinhole projector.

2024 Solar Eclipse: Celestial wonder, thrilling astronomers

The 2024 solar eclipse is a celestial event that is sure to thrill astronomers and anyone interested in space and the cosmos. This will be a total solar eclipse, which means that the moon will pass directly between the sun and the Earth, blocking out the sun's light and casting a shadow on the Earth.

This solar eclipse will be visible from North America, specifically Mexico, the United States, and Canada. People in these regions will be able to witness the moon completely obscuring the sun, creating a breathtaking sight. Observers in the path of the eclipse will experience a few minutes of darkness in the middle of the day, and the surrounding areas will experience a partial eclipse.

The 2024 solar eclipse is significant because it is the first total solar eclipse visible in North America in nearly four decades. It is also the first total solar eclipse to cross the contiguous United States since the historic eclipse of June 8, 1918. This has created a lot of excitement among astronomers, as well as amateur skywatchers and anyone interested in the cosmos.

Astronomers are eagerly anticipating the 2024 solar eclipse because it provides a unique opportunity to study the sun and the moon. During the eclipse, they will be able to gather important data on the sun's corona, which is the outer atmosphere of the sun. The corona is normally difficult to observe because it is overwhelmed by the bright light of the sun, but during a total solar eclipse, the moon blocks out the sun's light, allowing astronomers to study the corona in detail.

In a nutshell, the 2024 solar eclipse is a celestial wonder that is sure to thrill astronomers and anyone interested in space and the cosmos. With its path crossing North America, it provides a unique opportunity for people in Mexico, the United States, and Canada to witness a total solar eclipse, and for astronomers to study the sun and the moon in detail. Whether you're an astronomer or just a curious skywatcher, the 2024 solar eclipse is an event not to be missed.

Best Viewing Spots For 2024 Solar Eclipse

The 2024 total solar eclipse will be visible on April 8, 2024 and will be a unique opportunity for sky gazers to witness this celestial event. Here are some of the best viewing spots for the 2024 solar eclipse:

Mexico City, Mexico: The eclipse will first make landfall near Mexico City, making it an ideal spot for viewing the event. The city is easily accessible and offers a wide range of accommodations, making it a convenient option for those traveling to see the eclipse.

Texas Hill Country, United States: This region of Texas will offer some of the best views of the 2024 solar eclipse in the United States. The rolling hills and clear skies make it an ideal location for viewing this rare event.

Little Rock, Arkansas: Little Rock is one of the cities that will experience the longest duration of the total eclipse, making it a prime viewing spot. With its central location, it is easily accessible for people coming from other parts of the country.

Paducah, Kentucky: Paducah is another city that will experience a long duration of the total eclipse and offers clear skies, making it an excellent location for viewing the event.

Carbondale, Illinois: Carbondale is the point of greatest duration for the 2024 solar eclipse, meaning it will experience the longest period of darkness during the event. This city is also the site of the total solar eclipse that occurred in 2017, making it a popular destination for eclipse chasers.

Regardless of where you choose to view the 2024 solar eclipse, it is important to always use proper eye protection, such as eclipse glasses or a solar viewer, to avoid damage to your eyes.

How Improved Technology Will Enhance The Upcoming Eclipse Viewing Experience

With advancements in technology, the viewing experience of eclipses has improved dramatically in recent years. Here are a few examples of how technology has and will continue to enhance the experience:

(1.) Better viewing equipment: With the development of high-quality telescopes, binoculars, and cameras, it is now possible to observe and capture detailed images of eclipses with greater clarity. The increased magnification and resolution provided by these tools enhances the viewing experience, allowing people to see the intricate details of the eclipse in a way that was previously not possible.

(2.) *Virtual and augmented reality:* With the growth of virtual and augmented reality technology, it is now possible to experience an eclipse in a fully immersive way. Using VR and AR devices, viewers can not only see the eclipse in real-time but also learn about its various stages, as well as the science behind it.

(3.) *Streaming and live coverage:* The widespread adoption of high-speed internet and the increasing popularity of streaming services have made it possible for people to view live coverage of eclipses from the comfort of their own homes. This means that people no longer have to travel to remote locations to see an eclipse and can experience it from anywhere with a stable internet connection.

(4.) Predictive software: Predictive software and mobile apps have made it easier for people to plan for and observe eclipses. These tools provide detailed information about the timing and location of eclipses, as well as advice on how to safely view them.

Improved technology is playing a key role in enhancing the upcoming eclipse viewing experience. With better viewing equipment, virtual and augmented reality, live streaming, and predictive software, people can now observe eclipses in a more comfortable and informed way than ever before.

Types of Solar Eclipses and their differences from each other

Solar eclipses are a rare and awe-inspiring phenomenon where the Moon passes between the Earth and the Sun, blocking out the Sun's light and casting a shadow on Earth. There are three types of solar eclipses: total, partial, and annular.

(1.) Total Solar Eclipse:

This type of eclipse occurs when the Moon completely covers the Sun, allowing only the Sun's corona to be visible. This type of eclipse is only visible from a narrow path on Earth known as the path of totality. The last total solar eclipse occurred on December 14,

2020, and was visible from parts of South Africa, Chile, and Argentina.

(2.) Partial Solar Eclipse:

A partial solar eclipse occurs when the Moon only partially covers the Sun. This type of eclipse is visible from a larger area on Earth, but the Sun remains partially visible throughout the eclipse. The last partial solar eclipse occurred on June 10, 2021, and was visible from parts of Canada, Greenland, Russia, and northern China.

(3.) Annular Solar Eclipse:

An annular solar eclipse occurs when the Moon is at its farthest point from Earth and is not large enough to completely cover the Sun. This creates a ring of light around the Moon, making it look like a "ring of fire."

The last annular solar eclipse occurred on June 21, 2020, and was visible from parts of Africa, the Middle East, and Asia.

Unique Challenges Associated with Viewing of Solar Eclipses and How to Overcome Such

Inasmuch that Solar eclipses are a natural phenomenon that captivates and enthralls people from all over the world. However, viewing such scenario presents its own set of unique challenges that need to be overcome in order to make the most of the experience.

(1.) Eye safety: One of the biggest challenges associated with viewing solar eclipses is eye safety. Looking directly at the Sun can cause permanent damage to your eyes, even during a partial eclipse. It is important to use special protective glasses that are designed specifically for viewing solar eclipses.

(2.) Weather: Another challenge is the weather. Solar eclipses can be obstructed by cloud cover, which can diminish the viewing experience. If the weather forecast is not favorable, it is recommended to view the eclipse online or through live streaming.

(3.) Location: Finding the right location to view the solar eclipse can also be a challenge. Some areas may experience only a partial eclipse, while others will experience a total eclipse. To view a total eclipse, it is important to find a location within the path of totality.

(3.) Crowds: Crowds can also be a challenge when viewing solar eclipses. Large crowds can obstruct the view and make it difficult to find a clear spot to observe the eclipse. It is recommended to plan ahead and arrive early to secure a good viewing spot.

(4.) Technology: Technology can also pose a challenge when viewing solar eclipses. Using devices such as smartphones and cameras to view the eclipse can interfere with the viewing experience. It is recommended to use specialized viewing devices or filters to protect the eyes and improve the viewing experience.

To overcome these challenges, it is important to plan ahead and be prepared.

Here are a few tips on how to make the most of a solar eclipse viewing experience:

- Purchase solar-safe viewing glasses or special filters to protect your eyes.
- Check the weather forecast ahead of time and have a backup plan if the weather is not favorable.
- Arrive early to secure a good viewing spot.
- Use specialized viewing devices or filters to protect your eyes and improve the viewing experience.
- Respect the environment and leave the viewing area in its original state.

How can you capture the best photos and videos of the 2024 Solar Eclipse, and what gear should you bring with you?

The 2024 solar eclipse is a highly anticipated astronomical event, and capturing the best photos and videos requires careful planning and preparation. Here are some tips to help you get the best shots:

(A.) Location: Choose a location with a clear view of the sky and as far away from light pollution as possible. Consider the landscape and how you can incorporate it into your shots for added interest.

(B.) Camera Gear: For photography, a digital single-lens reflex (DSLR) camera is recommended for its manual controls and ability to change lenses. A telephoto lens with a focal length of 300mm or longer will allow you to zoom in on the eclipse and capture fine details. A tripod is also essential to keep your camera steady during long exposures.

(C.) Protective Filters: Never look directly at the sun without protective equipment, including certified solar filters for your camera and your eyes. Make sure your camera filter is specifically designed for solar eclipses and fits securely on your lens.

(D.) Settings: When taking photos, set your camera to manual mode and adjust the aperture, ISO, and shutter speed accordingly. Experiment with different settings to find what works best for you. During the partial phases of the eclipse, you can use a lower ISO and shutter speed, while during the total eclipse, you may need to increase the ISO and use a faster shutter speed to capture the detail in the corona.

(E.) Video Gear: For videography, consider using a mirrorless camera or a camcorder with manual controls and the ability to change lenses. A stabilizer, such as a gimbal, will help keep your shots steady and smooth.

(F.) Composition: In addition to capturing the eclipse itself, consider incorporating other elements into your shots, such as the landscape, people, or wildlife. Experiment with different angles and compositions to create unique and interesting images.

In summary, to capture the best photos and videos of the 2024 solar eclipse, you should bring:

- DSLR camera or mirrorless camera/camcorder

- Telephoto lens with a focal length of 300mm or longer

- Tripod

- Solar filter for your camera and eyes

- Stabilizer (gimbal) for videography

By following these tips and having the right gear, you'll be well on your way to capturing amazing photos and videos of this once-in-a-lifetime event!

Some Cultural and Historical Significance of Solar Eclipses and The various Responses of The Populace Towards Such Over The Years

Solar eclipses have been significant cultural and historical events since ancient times. They have been seen as omens or portents of change, and different civilizations have responded to them in various ways.

In ancient cultures such as the Babylonians and the ancient Chinese, solar eclipses were seen as omens of disaster, and the ruling monarchs would perform rituals to try to appease the gods and prevent the disaster.

In medieval Europe, solar eclipses were viewed with fear and superstition, and some believed that they signaled the end of the world. During these times, religious leaders would hold mass to pray for protection and deliver sermons on the importance of repentance.

In some Native American cultures, eclipses were viewed as opportunities to renew spiritual connections. They would hold rituals and offer offerings to the sun to ensure that it would return.

In modern times, solar eclipses are viewed as scientific events and are studied by astronomers, who use them to learn more about the sun and the solar system.

However, for many people, they still evoke a sense of awe and mystery, and draw large crowds of onlookers who are eager to witness the spectacle.

In conclusion, solar eclipses have been significant cultural and historical events for centuries, inspiring responses ranging from fear and superstition to scientific curiosity and wonder. Despite the advances in our understanding of these events, they continue to captivate people and evoke strong emotions.